HC-05 Bluetooth

Arduino

How to program it and use it
with your Arduino Projects.

(Includes instructions for the ZS-040)

Support Website:

www.Y2KLeader.com

Contents

Introduction

What is the HC-05?

The HC-05 is a very inexpensive ($3-$5) Bluetooth wireless serial control module that can be paired with a microcontroller such as the Arduino and used to communicate with another Bluetooth module or device such as a smart phone or tablet.

How this Book Can Help You

If you've come across this book, it's probably because you already bought an HC-05 or are interested in adding Bluetooth capability to your robotic projects. You've no doubt already discovered that it can be a little tricky. Have no fear – this book will take they mystery and frustration out of working with the HC-05 and the ZS-040.

HC-05 vs. HC-06

The HC-05 can be run as either a "Master" or "Slave" whereas the HC-06 can only be run as a "Slave". Also, the firmware is different on the two devices. The HC-05 has six pins and the HC-06 only has four pins. It doesn't use a state pin or a Key/En pin. On the HC-06 there is usually a label that says "Wakeup" but there is no pin attached. Since the HC-06 can only be used as a "Slave" then it only needs four pins. If you have a JY-MCU model it is probably an HC-06. Again, you can easily tell because the HC-06 has only 4 pins.

ZS-040

You may have ordered an HC-05 and received a ZS-040. Don't worry. A ZS-040 is an HC-05, but there are some things that you need to know in order to program it. I'll cover this topic in detail in t his book so there will be no more confusion and frustration.

Parts and Tools You Will Need

Each project in this book has its own list of parts needed for that specific activity. Most parts can be bought from Ebay.com or Amazon.com. You will save a lot of money if you buy the parts in bulk, meaning buying many at once. I will list the parts first, then I will go into detail on each part telling what it does and where to buy it.

- Arduino
- Arduino IDE Software (Runs on Windows, MAC or Linux)
- Power Supply for Arduino (9-12 Volt 1Amp)
- USB Cable
- Male-Male DuPont Cables (Jumper Wires)
- Male-Female DuPont Cables (Jumper Wires)
- Female-Female DuPont Cables (Jumper Wires)
- Resistors
- LED's
- 4-Relay Module
- MOSFET Transistors (FQP30N06L)
- Two HC-05 Bluetooth Modules
- Tablet or Smart Phone
- 22 Gauge Solid Core Wire

Tools Needed

- Wire strippers (If using 22 Gauge wire instead of jumper wires)
- Jeweler's flathead screwdriver
- Magnifying glass or darn good eyesight.

Connecting to the HC-05

USB to FTDI Converter
(www.Ebay.com)

Arduino

In order to program the HC-05, you must connect to its serial UART interface. I will show you two ways that you can use to connect to it so that you can program it. These are:

1. USB to FTDI Converter
2. Arduino

I'll first show you how to use all the methods listed above for the hardware interface to the HC-05 / ZS-040, then I'll show you how to load the Arduino's Serial Terminal for the software interface. Once your hardware and software interface is setup, I'll show you how to use the AT Commands to encode the Bluetooth modules with the proper commands depending upon the scenario for which you are using them for.

CAUTION:

USB to TTL Converter
Adapter Module
(www.Ebay.com)

There is another way that you may see recommended to use to connect directly to the HC-05, which is the USB to TTL Adapter Module (left). The one I used got VERY hot. It destroyed one of my HC-05 modules. I've read that others had the same problem so I do not recommend this type of unit. However, I have had good luck with the USB to FTDI Converter and the Arduino.

Option #1 – Connecting to the HC-05 with a USB to FTDI Converter

USB to FTDI Converter
(www.Ebay.com)

FTDI stands for "Future Technology Devices International", which is the company that developed the technology. It is not expensive ($4-$6 + the cable) and requires special drivers. You also need a special cable – a USB 2.0 Type A to USB Type 2.0 Mini B Cable like the one shown to the right. There has been

USB 2.0 Mini B

some problems and controversy regarding the drivers, which were reportedly made to keep the hardware proprietary. That being said, I have used this and it worked well once I installed the driver. The driver can be found at:

http://www.usb-drivers.org/ft232r-usb-uart-driver.html

The website gives a good tutorial on how to download and install the driver. Once the driver is installed you need to use your Serial Terminal application to connect with your device using the above-mentioned COM port, baud rate and connection options. I will explain more about the Serial Terminal applications later in this chapter.

First wire the adapter to the HC-05 ash shown here:

HC-05 TXD → RX
HC-05 RXD ← TX
HC-05 VCC + EN ←VCC
HC-05 GND ← GND

→ PC USB Port

USB to FTDI Converter Pinout

You can either use a breadboard to connect it or connect it directly. You won't need another power source for the HC-05 since the USB to FTDI Converter will supply 3.3 Volts of power. Don't forget to set the jumper on the USB to FTDI Converter to 3.3 Volts as shown here:

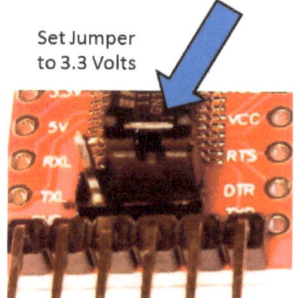

Set Jumper to 3.3 Volts

You can connect the USB to FTDI Converter to the HC-05 using a breadboard like this:

Important: Don't forget to connect the Vcc (+3.3 Volts) to the EN pin. The best way to do this is simply to connect an extra jumper wire from the Vcc pin to the EN pin. If you forget to do this you will only be able to send some AT commands but not all of them. I call it the "Baby AT" vs. the "Adult AT" mode. Here is a wiring diagram:

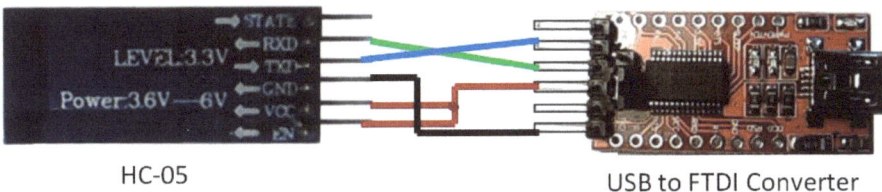

HC-05   USB to FTDI Converter

Now connect the USB cable from your PC's USB port to the USB to FTDI Converter, load your Serial Terminal program and you are ready to start typing AT Commands to program your HC-05 Bluetooth module using the Arduino IDE's serial terminal .

You should now be able to use Termite RS232 Communications too or the Arduino IDE Serial Terminal to send AT commands to the HC-05. I will explain more about how to connect using both of these software options in the chapter, "Programming the HC-05 and ZS-040".

Option #2 – Connecting to the HC-05 with the Arduino

If you don't have a USB to Serial TTL Converter, then you can easily program the HC-05 (and ZS-040) with the Arduino. The wiring is fairly easy with one consideration – you must step down the voltage from the pin on the Arduino that you will be using as the TX (Serial Transmit) line from 5 Volts to 3.3 Volts. If you don't, then you will eventually fry your Bluetooth module. This is accomplished with the use of a voltage divider or Logic Level Converter (See the chapter titled, "Voltage Dividers and Logic Level Converters").

Here is the wiring diagram, followed by a detailed wiring description and the HC-05 Programmer Sketch.

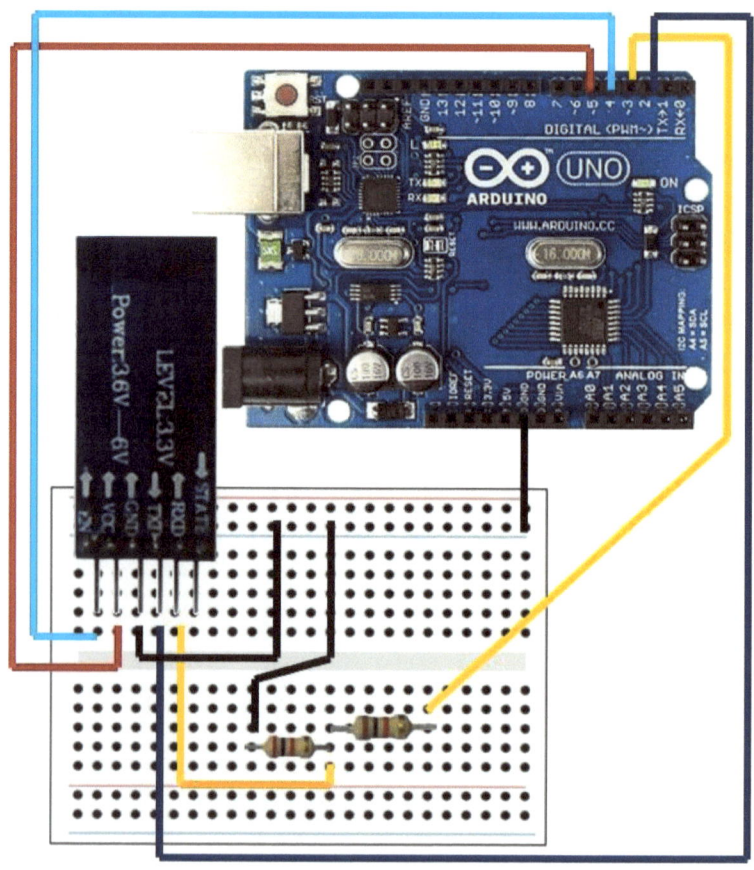

1. Connect the GND pin on the Arduino to the common ground rail on the breadboard.
2. Connect Pin 3 on the Arduino to one end of a 2,000 Ohm resistor (see "Voltage Dividers and Logic Level Converters" for an explanation).
3. Connect the other end of the 2000 Ohm resistor to a 1000 Ohm resistor. At the junction between the two resistors, connect a wire to the RXD pin on the HC-05 Module.
4. Connect the other end of the 1000 Ohm resistor to the common ground rail.
5. Connect a jumper from the common ground rail to GND on the HC-05.
6. Connect a jumper from pin 2 on the Arduino to the TXD pin on the HC-05.
7. Connect Pin 4 on the Arduino to the EN (sometimes labeled KEY or WAKEUP) pin on the HC-05.
8. Connect Pin 5 on the Arduino to the VCC pin on the HC-05.

Note:
In case you were wondering why she STATE pin wasn't connected (or what the state pin is) – The STATE pin outputs a signal of zero or 1 (LOW or HIGH). It is LOW if there is no Bluetooth signal and HIGH if there is one. It is not needed here, but it might be useful if you would read the value of the STATE pin every so often to see if there was a signal. If there is no signal, you could stop your vehicle or whatever you are trying to control or program it to return home, provided it knows where home is.

HC-05 Programmer Arduino Sketch

```
/*
HC-05 AT Programmer by Michael Wright (y2kLeader.com)
HC-05 TXD - Arduino Pin 2
HC-05 RXD -  Voltage Divider or LLC - Arudino Pin 3
HC-05 EN - Arduino Pin 4
HC-05 VCC - Arduino Pin 5
HC-05 GND - Arduino GND
Voltage divider 1000 Ohm and 2000 Ohm resistor
If using a ZX-040 connect 3.3 Volts to pin 34 &
  hold button when powering on to enter AT mode.
*/
#include <SoftwareSerial.h>
int rx=2;
int tx=3;
int enable=4;
int power=5;
int data;
SoftwareSerial bluetooth(rx,tx);
void setup() {
  Serial.begin(38400);
  pinMode(enable,OUTPUT);
  pinMode(power,OUTPUT);
  digitalWrite(enable,HIGH);
  bluetooth.begin(38400);
  digitalWrite(power,HIGH);
  Serial.println("Ready");
}
void loop() {
  if (bluetooth.available()){
    Serial.write(bluetooth.read());
  }
  if (Serial.available()){
    data=Serial.read();
    bluetooth.write(data);
  }
}
```

You should now be able to use the Arduino IDE Serial Terminal to send AT commands to the HC-05. I will explain more about how to connect using both of these software options in the chapter, "Programming the HC-05 and ZS-040".

IMPORTANT NOTE:

The "HC-05 Programmer" sketch is used for "PROGRAMMING" the module, not "RUNNING" it. When you go to run the module after being paired with another Bluetooth, you will have to upload a sketch that reads and/or writes data from/to the Module. When it reads data, the Arduino sketch that has been uploaded will respond to the incoming commands either by performing a pre-programmed task (such as turning a motor or servo) or by sending information the other module. See the chapter "Pairing Scenarios".

Connecting to the ZS-040

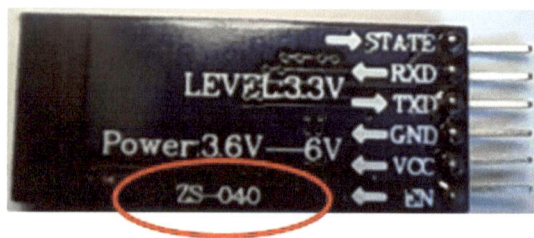

The ZS-040 model has a quirk. Unlike the regular HC-05, the ZS-040 will not go into full AT mode unless you bring pin 34 up to 3.3 Volts. On both the HC-05 and the ZS-040 you must bring the EN (sometimes labeled KEY) pin high by supplying 3.3 or 5 Volts to it. However, with the ZS-040 you also have to bring Pin 34 high. You can do this by soldering a wire (see picture to the right) onto the pin or using an alligator clip.

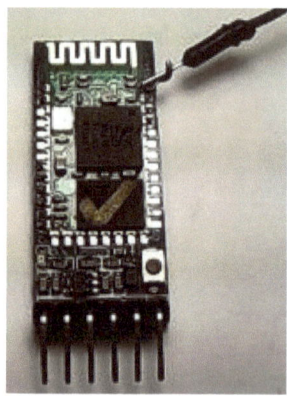

If you have a very tiny alligator clip it works best, but I've been able to do it with a regular sized alligator clip. Pin 34 is the last pin on the top-right side of the module. Of

Place an alligator clip on pin 34 that is at 3.3 Volts if you have an SZ-040 model.

course be very careful. You don't want to damage your module by scratching it or touching the wrong pin and shorting it out. I have been able to do this successfully, but after about 10 times connecting it, I couldn't get the alligator clip to make good contact anymore.

Once Pin 34 is connected to, then the other end of the jumper wire or alligator clip can be inserted into the 3.3 or 5 Volt pin on the Arduino or the breadboard rail. The alligator clip has top and bottom teeth.

The top tooth should go onto the top of the pin so that it makes good contact as shown in the picture here:

Place this tooth on top of the pin.

Also, with the ZS-040, **you have to hold down the button** on the module and either reset the Arduino or pull the ground or power jumper from the breadboard and re-insert it when powering-on. This is the only way that you can get the ZS-040 to go into AT mode. The HC-05 only needs the EN/KEY pin to be set to HIGH but the ZS-040 requires that that button on the module be pressed and held before power is supplied. Yes, it's a real pain with the ZS-040, but if you already have it, then this information is what you need to get it programmed.

NOTE:

The ZS-040 model tends to easily have a problem with holding the connection. If your particular module can't keep a connection for more than a few seconds, it may be a defective module. You can sometimes fix the problem by making sure not to have the USB cable connected between your PC and the Arduino. Honestly, it's best to stick with the standard HC-05 modules rather than the ZS-040.

Programming the HC-05 and ZS-040

Now that you are either connected to your HC-05 or ZS-040 with the USB to Serial adapter, USB TTL converter or the Arduino, you are ready to start programming.

If you are using the Arduino to program it, then you must upload the sketch "HC-05 Programmer" to your Arduino. Once it is uploaded and running, you should see the light start blinking on your module. If there is no light on your module, check your connections. You can now click on the Serial Terminal 🔍 on the upper-right side of your Arduino IDE to bring up the serial monitor.

If you are using the USB to Serial adapter or USB TTL converter to connect to your module, then the lights should come on automatically. If they don't then check your connections. You now have two choices of software to use to connect to it. You can use the Arduino IDE's Serial Terminal or you can use Termite (See "Installing and Using Termite").

Whether you are using the Arduino's Serial Terminal (I call this Terminal but some people call it the Serial Monitor), Termite or another RS232 Serial connection program, you must set the baud rate to 38,400, 8 bits, 1 stop bit and no parity. If the wrong baud rate is set, you will see either gibberish or nothing at all appear on your Serial Terminal's response area.

Now that your module is connected, your Arduino HC-05 Programmer Sketch is running (or Termite is running) and you are in AT programming mode the light on your module should be blinking **two seconds on – two seconds off.** If so, then you are ready to start entering AT commands into the Arduino Serial Terminal. If the light is flashing quickly it means that you are not in AT programming mode. This most often happens with the ZS-040. Just hold the button down on the module then reboot the power to the module. If you are not in AT mode on the regular HC-05 module, check to make sure that here is power going to the EN pin.

When first connecting to the HC-05 it normally has a default baud rate of **38400**, but if it won't connect then try 9600.

Here is a list of what the flashing lights mean on the module:

- When the module goes into AT command mode, the light will blink at two second intervals – on for two seconds, then off for two seconds.

- When it is searching or waiting for a connection to another Bluetooth module it will blink very fast continuously.

- When connected to another Bluetooth module, it will blink once per second.

The HC-05, if left alone in its freshly unboxed state, can be used as "Slave". The default device name is "**HC-05**" and the default password is "**1234**". If you are using it as the only Bluetooth device or if you know it will not be in an area in the future that has other Bluetooth devices, then you can just leave it alone and link to it with an Android phone, tablet or other HC-05 Module that is in the "Master" mode using the default credentials. If however, you want to change the device name and password then you will need to reprogram the HC-05 by sending it commands called "AT" commands.

Using the Arduino Serial Terminal

The Arduino IDE has a very easy-to-use Serial Terminal. It works with Windows, Linux and Mac, so it's universal. You don't need an Arduino attached to use the Serial terminal, but you must attach the USB to Serial Device. Load the Arduino IDE, then simply

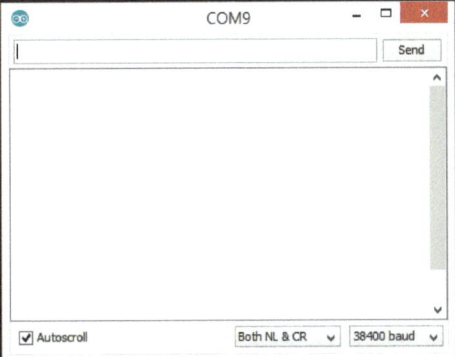

click on the 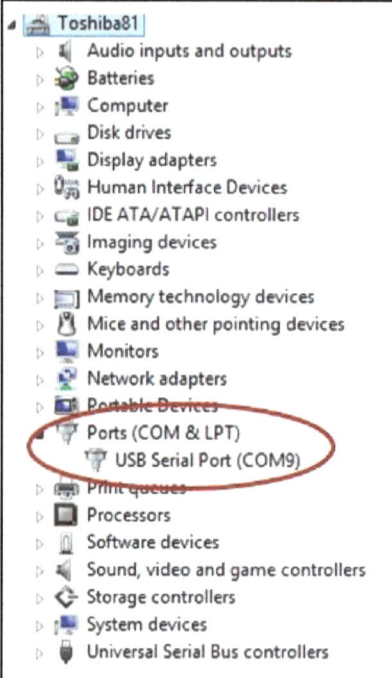 in the upper-right corner of the IDE window and the Serial Terminal window will appear.

Make sure that the baud rate is set for **38400 baud**. Also don't forget that you have to **append a carriage return and line feed** at the end of each line.

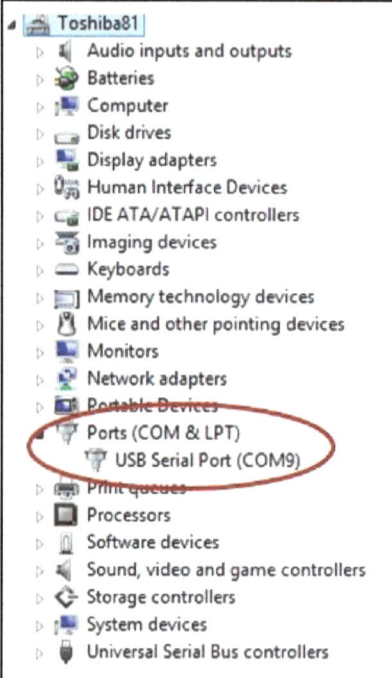

If you have difficulty finding the correct COM port once the cable is connected to the converter/adapter, go into Windows Control Panel and select the Device Manager. In the hardware tree, choose "Ports (COM & LPT)" and expand the tree. You should see "USB Serial Port (COM9) or whatever COM port number it assigned. If you don't see this then there is a problem with the serial port driver for windows or for your device (if any). Most of the time the device is automatically detected by Windows, Linux and Mac.

As you can see, the serial port was found on COM9.

Once you load the Arduino IDE, choose "Tools" from the menu and you should see Port: "Com9" in the drop-down menu. As shown below.

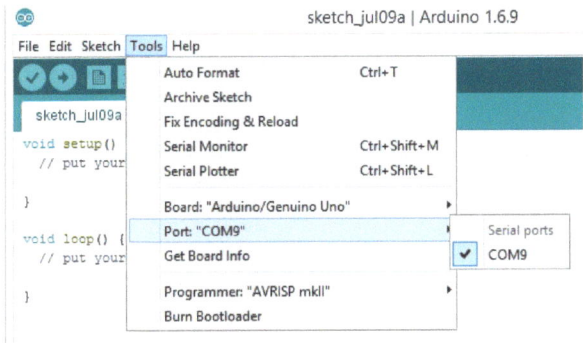

Make sure to choose COM9 from the secondary drop down list box and then you should be able to load the Serial Terminal with no problem.

Now it's time to start programming the HC-05.

IMPORTANT:
Right now you should either have the HC-05 connected to a USB to Serial TTL Cable or adapter that is attached to your PC **OR** have the HC-05 connected to the Arduino (usually using a solderless breadboard). If us are using the method with the Arduino then you MUST have the HC-05 Programmer Sketch uploaded into the Arduino. If you don't then the Arduino Serial Terminal won't give you a response. If there is no Arduino but you are attached to the HC-05 with the USB to Serial TTL cable or adapter, then you can use the Arduino IDE's Serial Terminal directly without having a sketch uploaded.

When you start the Serial monitor, it should say, "Ready". If it doesn't or there is gibberish on the screen, then check to make sure your baud rate is 38400 and that you have "Both NL & CR" in the box next to the baud rate.

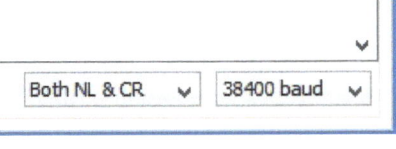

You should now see "Ready" like this:

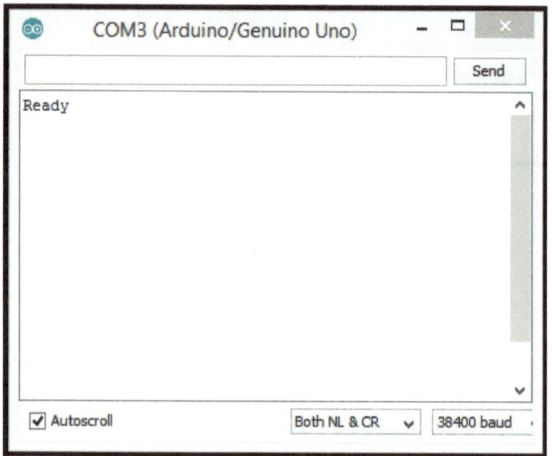

Now type, "AT" (without the quotes) in the text box on top. When you press the <Enter> key, it should respond with "OK".

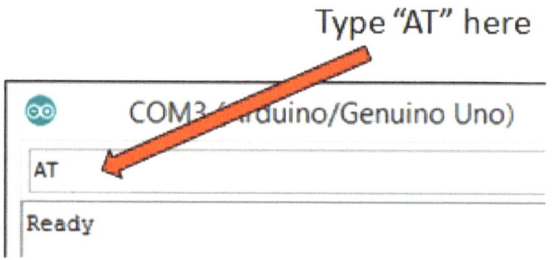

After you press <Enter> the "AT" that you just typed will go away and you should see "OK" like so...

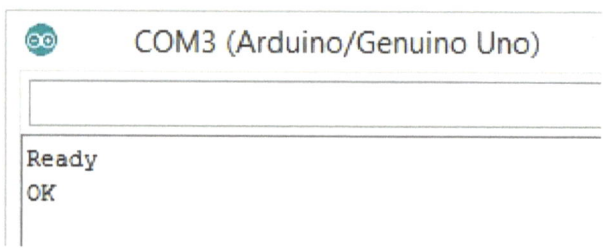

Now you are set. You can start entering the AT commands so that you can start programming them to Pair with each other or other Bluetooth devices. If it ever stops responding, check the LED on the module to make sure it is still in AT command mode by blinking in a long two second loop.

Next I'll show you how to program a module as a SLAVE and a MASTER.

Programming a SLAVE

The HC-05's out-of-box state is that of a SLAVE. The default password is 1234. I'll show you how to change this if you need to.

AT+ORGL //set to factory original default settings
AT+ROLE=0 //set to slave mode (if not at default state)
AT+NAME=MYSLAVE //name the module
AT+PSWD=1234 //set the password
AT+ADDR //retrieves the address (write this down for later)

Now the module is ready to be connected to an Arduino that has a sketch running on it that is customized for the slave.

Programming a MASTER

AT+ORGL //set to factory original default settings
AT+ROLE=1 //set to master mode
AT+NAME=MYMASTER //name the device (optional)
AT+PSWD=1234 //set password to the same as the slave device
AT+INIT //start the SPP (Serial Port Profile)
AT+INQ // search and display a list of remote devices
AT+PAIR= 2016,6,61551,10 //pairs with selected device from list
AT+BIND= 2016,6,61551 //binds with selected device
AT+LINK= 2016,6,61551

Now the module is ready to be connected to an Arduino that has a sketch running on it that is customized for the master.

Pairing Scenarios

There are different scenarios that you can use for your module(s). Each scenario will require different AT commands to be sent to the module(s). You can program these modules one at a time, but I find it useful to connect one module to the Arduino and connect another one either to a second Arduino or to a USB to Serial TTL adapter.

1. **Pairing an HC-05 Slave to an Android Master**
   Set your module as a Slave and pair it with a controller app on your smart phone or tablet. The slave module that is attached to an Arduino can relay signals to your robotic vehicle, drone, etc.

2. **Pairing an HC-05 Master to an HC-05 Slave**- Set one module as a Master and the other one as a slave. Either the Master or the Slave can be a custom built joystick and the other one your robot. You will need two Arduinos for this project.

Pairing an HC-05 Slave to an Android Master

SLAVE   MASTER

In this scenario I'll show you how to use an Android cell phone or tablet to connect to an HC-05 to turn a servo motor. First I'll show you how to write a sketch for this project. Then I'll show you how to wire it. Finally I'll show you how to use the Android to control it.

First, download the Arduino Bluetooth Controller from the Google Play store. You can also try other controllers, but this is one that I found that works well.

IMPORTANT:
Before uploading the "HC-05 Slave to Android Master" sketch, disconnect the RXD and TXD wires from the HC-05 Module or disconnect Pins 2 & 3 from the Arduino (or whatever pins you have assigned to the softwareserial). If you don't it will cause a conflict and your sketch may not upload. If for some reason you forgot to disconnect the pins and your sketch won't upload, go ahead and disconnect the pins and remove the serial cable from the Arduino and re-insert it.

Wiring Diagram

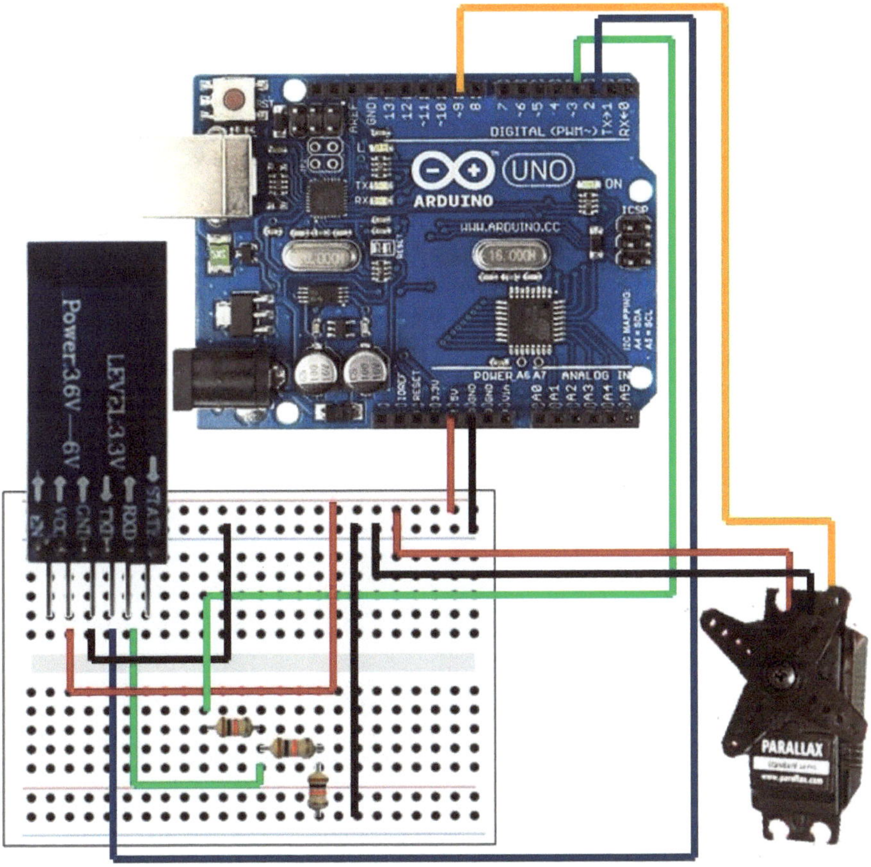

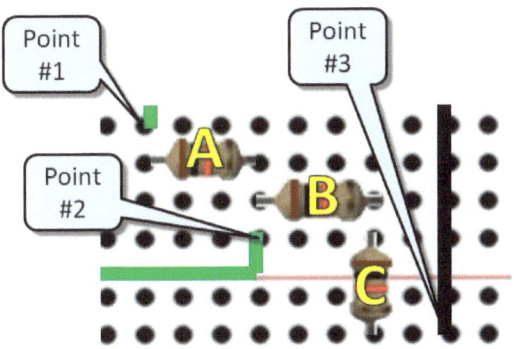

Voltage divider made with three 1000 Ohm resistors.

Wiring Directions

The "single" servo method can be powered with the power supply from either the PC to Arduino USB cable, the Arduino Power Jack or from the Vin pin on the Arduino. The Power Jack is the preferred method unless you have a regulated power supply that can supply power to the Vin pin on the Arduino. The Power Jack has a regulator already, whereas the Vin pin does not.

1. Connect Pin 2 of the Arduino to the TXD Pin of the HC-05.
2. Connect Pin 3 of the Arduino to Point #1 of the voltage divider circuit.
3. Connect Point #2 of the voltage divider circuit to the RXD pin of the HC-05.
4. Connect Point #3 of the voltage divider circuit to the ground rail.
5. Connect GND of the HC-05 to the Ground Rail.
6. Connect VCC of the HC-05 to the Power Rail.
7. Connect the Power rail to the 5V pin on the Arduino.
8. Connect the Ground rail to the GND on the Arduino.
9. Connect the ground wire of the servo to the ground rail.
10. Connect the power wire of the servo to the power rail.
11. Connect the data wire of the servo to Pin 9 on the Arduino.

NOTES:

Although you can use the Arduino to power a single servo, it is not the best practice because you are maxing out the amperage that the Arduino can handle. Also, you are powering the HC-05. If you will be putting load on the servo, I would definitely recommend using a separate power source. See the "Wiring Diagram (Multiple Servos)" section after this one.

HC-05 Slave to Android Master Sketch

```
//HC-05 to Android Master (Single Servo)
#include <Servo.h>
Servo servo1;
int servo1_position=0;
#include <SoftwareSerial.h>
int rx=2;
int tx=3;
int enable=4;
int power=6;
byte data;
SoftwareSerial bluetooth(rx,tx);
void setup() {
  servo1.attach(9);
  Serial.begin(38400);
  pinMode(enable,OUTPUT);
  pinMode(power,OUTPUT);
  digitalWrite(enable,HIGH);
  bluetooth.begin(38400);
  digitalWrite(power,HIGH);
  Serial.println("Ready");
}
void loop() {
  if (bluetooth.available()){
    data=bluetooth.read();
    Serial.write(data);
    if (data=='l'){
     servo1_position=0;
     servo1.write(servo1_position);
    }
    if (data=='u'){
     servo1_position=90;
     servo1.write(servo1_position);
    }
    if (data=='r'){
     servo1_position=180;
     servo1.write(servo1_position);
```

```
    }
  }
  if (Serial.available()){
    data=Serial.read();
    bluetooth.write(data);
  }
}
//end sketch
```

Now, when you connect the HC-05 slave to the Arduino and run the Arduino Bluetooth Controller, you can program the buttons on the App. I program the left button to send an "l" (lower case "L"), the top button to send a "u" and the right button to send an "r".

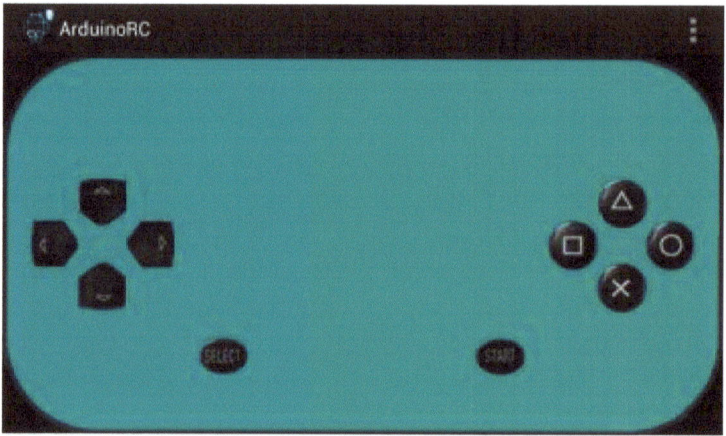

Arduino Bluetooth Controller
(www.play.google.com)

When the buttons are pressed your servo will position itself at 0, 90 or 180 degrees.

Pairing an HC-05 Master to an HC-05 Slave

Master   Slave

With this scenario you can build your own Bluetooth controller such as your own customized joystick and link it to your robot. You will need two Arduinos, one for the Master and one for the Slave.

When pairing two Bluetooth devices, one must be a "Master" and one much be a "Slave". You cannot have two masters or two slaves.

Step #1 – Create the "SLAVE" Module

If you are unboxing an HC-05, HC-06 or ZS-040 then the it should already be setup as a slave with the default password of "1234". If you are satisfied with this configuration, then you don't need to program it. If however, you wish to change the settings, continue with this step.

First, attach the HC-05 "Slave" Module and Enter into AT Programming mode.

If you are using a USB to Serial cable, load the communication software onto your PC then attach the HC-05 Module to the USB to Serial Cable. You will have to hold down on the button on the HC-05 while you plug it into the Serial module in order to enter into AT command mode.

If you using the Arduino to interface with the module then insert the module in the preset space on the breadboard. Make sure that the **"HC-05 Programmer"** sketch is loaded into the Arduino IDE. Send the sketch to the Arduino so that it starts running.

The HC-05 should now start blinking for two seconds on, then two seconds off continuously. If for some reason the HC-05 blinks twice quickly and waits for 2 seconds then blinks twice quickly and so on in a loop then it means that it is already paired with something else. If this happens, hold down on the button on the HC-05 and press the reset button on the Arduino. It should then blink two seconds on then 2 seconds off continuously.

If your HC-05 is the ZS-040 model, don't forget to attach a wire or alligator clip to pin 34 from the 3.3 Volt power rail. If you forget to do this, you will only be able to enter some AT commands but not the important ones. Also the ZS-040 model wants you to hold down the button on the module when powering it on.

Type only the command in bold faced print, pressing the <Enter> key at the end of each command:

AT+ORGL   //Resets the module to its factory state
AT+UART=38400,0,0 //set the baud rate to 38400 if not
AT+NAME   //write down the name it responds with
AT+NAME=SLAVE   //optional to give it a new name. "SLAVE" is the name I have mine for the example. You can replace this with anything you like.

AT+ADDR   //write down the address it responds with
AT+PSWD=1234   //set the password (default is 1234, 0000 is no passord)
AT+ROLE=0   //set for slave mode

The Slave module is now ready. Remove the Slave module from the breadboard or your USB to Serial Cable. It is now ready to be inserted into the breadboard and used as a Slave. Refer the the section in this chapter titled "Wiring and Connecting the Slave". You should have it powered on and ready to be paired. It should be blinking very fast, which means that it is searching for a Master device to pair with.

Here is the circuit layout and the Arduino sketch of the "Slave":

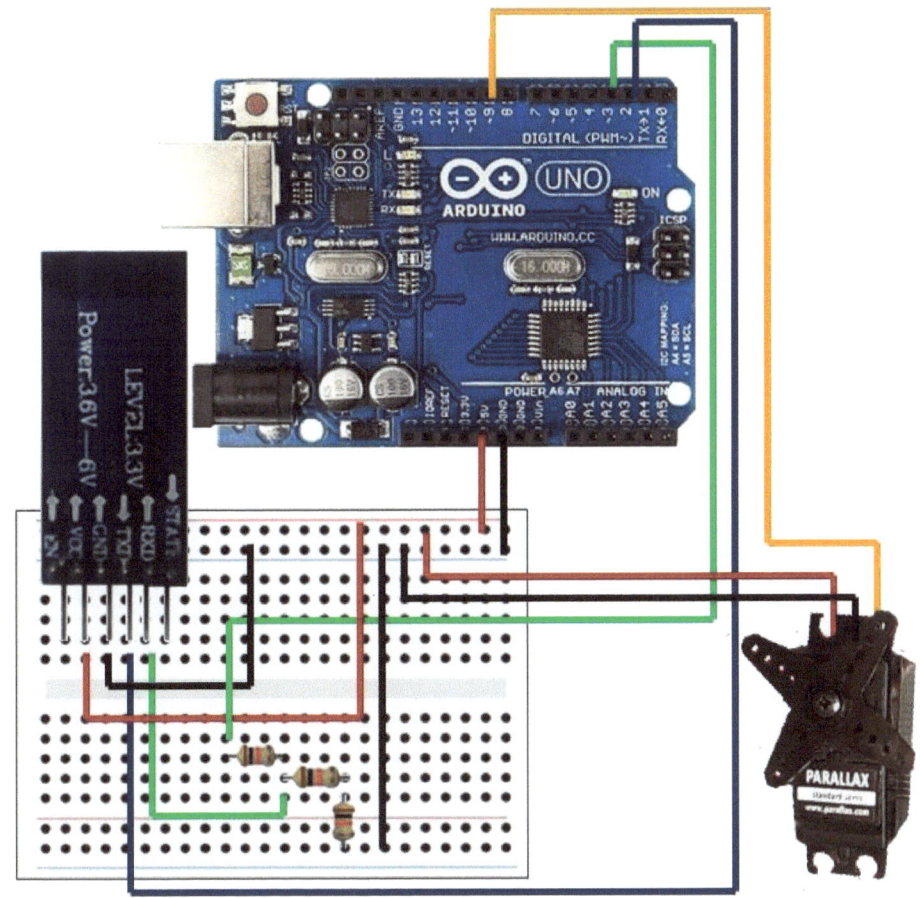

Slave Module & Arduino

Slave Wiring Directions

1. Connect Pin 2 of the Arduino to the TXD Pin of the HC-05.
2. Connect Pin 3 of the Arduino to Point #1 of the voltage divider circuit.
3. Connect Point #2 of the voltage divider circuit to the RXD pin of the HC-05.

4. Connect Point #3 of the voltage divider circuit to the ground rail.
5. Connect GND of the HC-05 to the Ground Rail.
6. Connect VCC of the HC-05 to the Power Rail.
7. Connect the Power rail to the 5V pin on the Arduino.
8. Connect the Ground rail to the GND on the Arduino.
9. Connect the ground wire of the servo to the ground rail.
10. Connect the power wire of the servo to the power rail.
11. Connect the data wire of the servo to Pin 9 on the Arduino.

"Slave" Arduio Sketch

```
//SLAVE
#include <SoftwareSerial.h>
#include <Servo.h>
Servo servo1;
int rx=2;
int tx=3;
int data_from_master=0;
SoftwareSerial bluetooth(rx,tx);
void setup() {
  Serial.begin(9600);
  servo1.attach(9);
  bluetooth.begin(9600);
  Serial.println("Slave Ready");
}
void loop() {
  if (bluetooth.available()){
    data_from_master=bluetooth.read();
    servo1.write(data_from_master);
    Serial.println(data_from_master);
  }
}
```

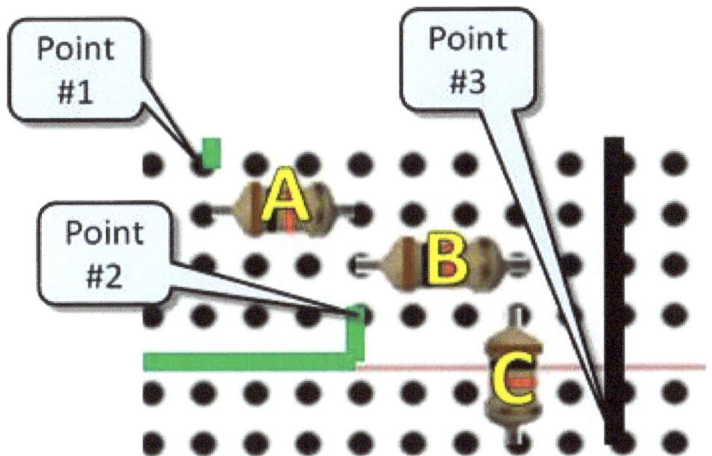

Voltage divider made with three 1000 Ohm resistors.

The voltage divider circuit above is necessary to convert the 5 Volt signal coming from the TX out pin on the Arduino to 3.3 Volts that the HC-05 Module can handle. If you do not use this voltage divider, then the 5 Volt signal will eventually damage your Bluetooth module. The Module can use 5 Volts for the VCC input but needs 3.3 volts for the UART serial interface.

See the chapter titled, "Voltage Dividers and Logic Level Converers" for more information about making voltage dividers.

Step #2 - Create the Master Module

Place the HC-05 module that you want to use as the Master into the breadboard and the Arduino with the "HC-05 Programmer" sketch loaded, or attach the HC-05 module to the USB to Serial Cable. Make sure that the module is blinking for two seconds on, then two seconds off to indicate that it is in AT programming mode.

Type only the command in bold faced print, pressing the <Enter> key at the end of each command:

AT+ORGL //Resets the module to its factory state
AT+RMAAD //clear any previously paired devices
AT+UART //displays the baud rate
AT+UART=38400,0,0 //set the baud rate to 38400 if not
AT+NAME=MASTER //optional to give it a new name.
AT+PSWD=1234 //set the password to the same as the slave

AT+ROLE=1 //set for master mode
AT+CMODE=1 //Set to be able to pair to any device
AT+INIT //Start serial port profile
AT+INQ   //Search for the slave

You will get a response that will list all the slaves found. It looks something like this:

```
+INQ:98D3:33:806A7D,0,7FFF
OK
```

The above should be the same address as the slave module from step #1. If it is not or nothing appears, make sure that the slave module is powered up and blinking quickly. This is telling you that the device address is "98D3:33:806A7D". However when you type this address later, you will replace the colons with commas.

If there is more than one module being seen or if you want to check the name, replace the colons with commas and type the following command like so:

AT+RNAME?98D3,33,806A7D

It will respond with the name given to the slave.

```
+RNAME:SLAVE
OK
```

Now pair it:

AT+PAIR=98D3,33,806A7D,30 //get ready to pair & wait
for 30 seconds (or whatever time you choose).

AT+BIND=98D3,33,806A7D //remember the connection
AT+LINK=98D3,33,806A7D //initiate the pairing

Now the Master and slave module should start blinking
twice fast, waiting for two seconds and twice fast
again continuously. This means they are linked.

AT+CMODE=0 //enter this command if you want to lock it
so that it only pairs with the specified slave that you
already paired with. If you do this, you can reset it
again later so that you can pair with others by typing:

AT+RMAAD
AT+CMODE=1
AT+RESET

Here is the circuit layout for the Arduino sketch for the "Master":

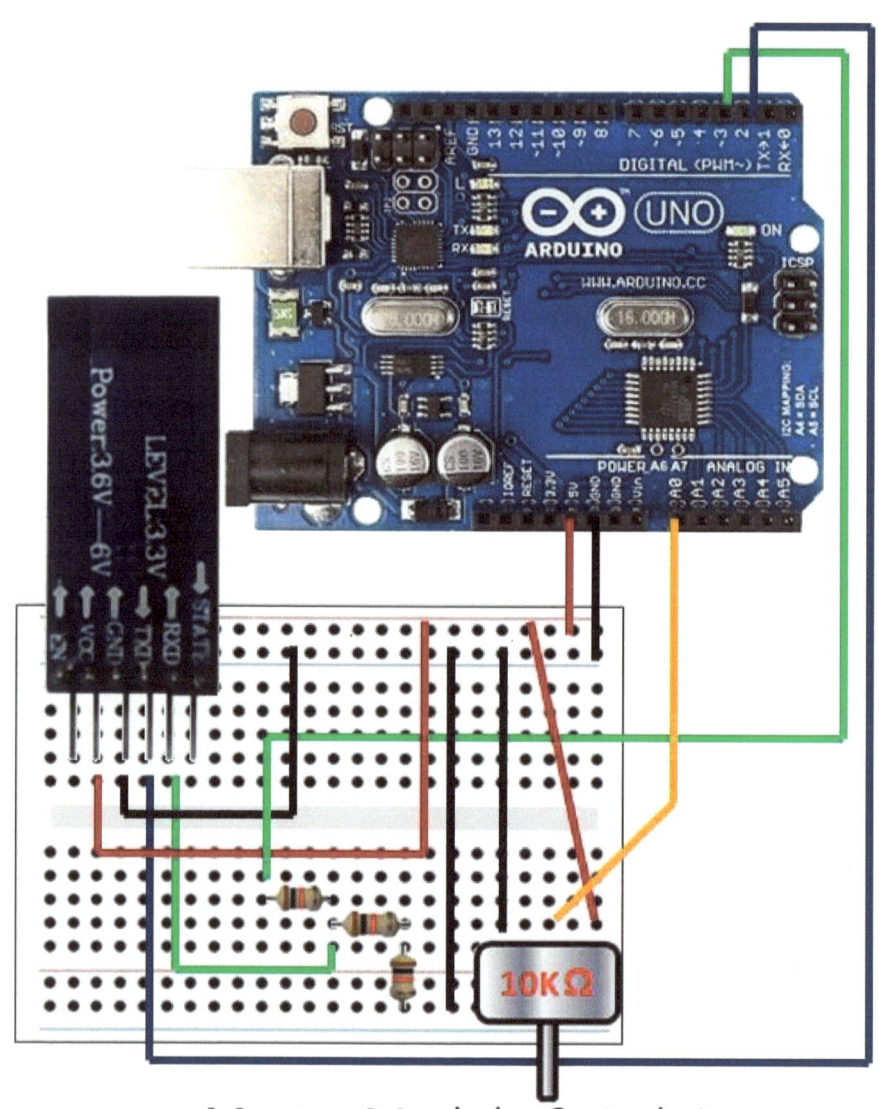

Master Module & Arduino

Master Wiring Directions

1. Connect Pin 2 of the Arduino to the TXD Pin of the HC-05.
2. Connect Pin 3 of the Arduino to Point #1 of the voltage divider circuit. (Refer to Slave diagram voltage divider)
3. Connect Point #2 of the voltage divider circuit to the RXD pin of the HC-05.
4. Connect Point #3 of the voltage divider circuit to the ground rail.
5. Connect GND of the HC-05 to the Ground Rail.
6. Connect VCC of the HC-05 to the Power Rail.
7. Connect the Power rail to the 5V pin on the Arduino.
8. Connect the Ground rail to the GND on the Arduino.
9. Connect the left lead of the potentiometer (pot) to the ground rail.
10. Connect the right lead of the pot to the 5 Volt power rail.
11. Connect the center lead of the pot to the A0 Pin (Analog zero) on the Arduino.

"Master" Arduino Sketch

```
//MASTER
#include <SoftwareSerial.h>
int rx=2;
int tx=3;
int pot=0;
SoftwareSerial bluetooth(rx,tx);
void setup() {
  Serial.begin(38400);
  bluetooth.begin(38400);
  Serial.println("Master Ready");
}
void loop() {
 pot = analogRead(A0);
 int potvalue = map(pot, 0, 1023, 0, 255);
 bluetooth.write(potvalue);//Sends pot value to slave
delay(10);
}
```

Now when you power on the Arduino (both master and slave) and turn the potentiometer knob, you will see the servo motor turn in unison with your potentiometer. Remember that the servo can turn 180 degrees. You may experience twitching on the servo. This happens because of interference in the data wire both from the potentiometer and to the servo. This may be resolved with a ferrite core, capacitors and/or software code. Check the support website for possible solutions.

I'll be programming a fully functional controller as part of an upcoming "Introduction to Robotics" book. It will include advanced Arduino sketch code for multiple Master inputs as well as receiving feedback from the Slave.

NOTES:

You may have noticed that the baud rates on the Slave are 9600 and the baud rates on the Master are 38400. I've found this to be the optimal settings for this configuration.

Voltage Dividers & Logic Level Converters

The HC-05 Module accepts 3.6 Volts - 6 Volts for its power supply needs.. However, the chip that runs the UART (Universal Asynchronous Receiver/Transmitter) uses 3.3 Volts to transmit and receive data over the RX (receive) and TX (transmit) and pins. This communication is not going out through the antennae, it is going back and forth from the Arduino.

The Arduino uses 5 Volts on its Serial RX and TX pins. The RX and TX pins on the Arduino are pin 0 and 1 respectively. However, when the Arduino is using the serial terminal for communication between the Arduino and the computer, it cannot use it for communication with another device such as the HC-05. This is why the softwareserial functions are used. Softwareserial allows the Arduino to assign other pins as an extra serial port in addition to pins 0 and 1. In the Arduino code in the previous chapters I assigned pins 2 and 3 as the RX and TX pins for the softwareserial communications.

Using Resistors as a Voltage Divider

Back to the voltage dividers – Since the Arduino outputs a 5 Volt signal through its pins, the TX (transmit) signal from pin 3 has too high of a voltage for the HC-05 to handle because can only handle 3.3 volts. If you send a 5 Volt signal directly to the HC-05, you will overload it and it will eventually burn up and die. You need to find a way to decrease the 5 Volt signal to 3.3 Volts. Most forums, websites and YouTube channels advise using a voltage divider consisting of two resistors to decrease the voltage. This is what is called a **"Voltage Divider"**. They advise using a 1000 Ohm and 2000 Ohm resistor to drop the voltage from 5 Volts to 3.3 Volts.

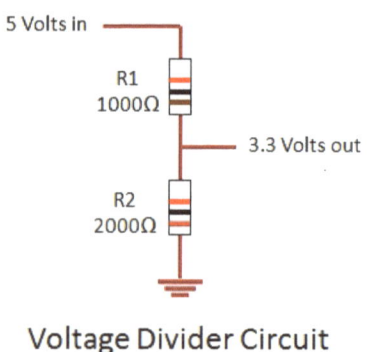

5 Volts in

R1
1000Ω

3.3 Volts out

R2
2000Ω

Voltage Divider Circuit

This will work for programming the HC-05 because it works at low speeds like 9600 baud. However there are a couple problems with this method. First of all, when you are paired with another Bluetooth device you may receive a stream of 0's when not receiving data. This is because the signal is being brought all the way to ground between data transfers. The other problem is that the resistors slow down the

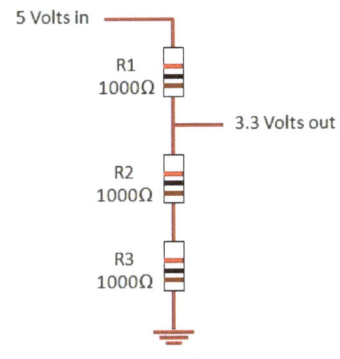

Voltage Divider Circuit

data transfer. This means that you may not receive data fast enough into your Arduino program to make your robot respond quickly enough. This could mean death for your battle bot or a crash and burn for your drone. If you don't have a 2000 Ohm (2K Ω) resistor you can link two 1000 Ohm (1000 Ω) resistors in a series like shown here:

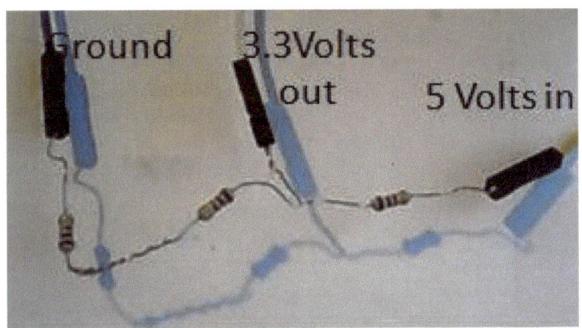

Actually the output is slightly lower (2.8-3.2 Volts). Mathematically it works out but there is always a little bit of voltage loss because of the resistance in the wires and breadboard connections.

Sizing Resistors for Other Power Levels

If you need to drop from let's say 9 volts to 3.3 Volts you would need different resistors. Generally resistors used for these low powered circuits range from 220Ω - 10,000Ω at ¼ Watts rating. There is a formula that you can use to figure out which resistors to use but it is very difficult to get the exact output voltage that you are looking for, but getting close is often good enough. You have to know :

1. The input voltage
2. The desired Output Voltage
3. The value of R1 (the first resistor leading from the input voltage).

First I'll show you how I figured out how to convert 5 Volts to 3.3 Volts, then I'll apply this to a 9 volt power source.

Pick a resistor size for R1 that you think will work, say 1000Ω. Use the following formula to computer for R2.

$$R_2 = R_1 \cdot \frac{1}{\left(\dfrac{V_{in}}{V_{out}} - 1\right)}$$

$$R_2 = 1000 \cdot \frac{1}{\left(\dfrac{5}{3.3} - 1\right)}$$

$$R_2 = 1000 \cdot \frac{1}{.515}$$

$$R_2 = 1000 \cdot 1.94$$

$$R_2 = 1941.7$$

So the value of the second resistor (R2) should be a resistor that is made that is close to 1941.7Ω. In this case it's a **2000Ω** resistor.

Let's say the input voltage is 9 Volts instead of 5 Volts. We can apply the same formula.

As you can see, the resistor size needed has to be close to 191.4. Looking on the resistor sizes chart in the back of the book, you can see that we can use either 180Ω or 200Ω resistor. A 180Ω resistor would bring the voltage slightly lower than 3.3 Volts and a 200Ω would bring it slightly higher.

$$R_2 = R_1 \cdot \frac{1}{\left(\dfrac{V_{in}}{V_{out}} - 1 \right)}$$

a

$$R_2 = 330 \cdot \frac{1}{\left(\dfrac{9}{3.3} - 1 \right)}$$

$$R_2 = 330 \cdot \frac{1}{1.7}$$

I've used this method without problems, even up to 38,400 baud.
Here is how it would look on a breadboard:

$$R_2 = 330 \cdot .58$$

$$\boxed{R_2 = 191.4}$$

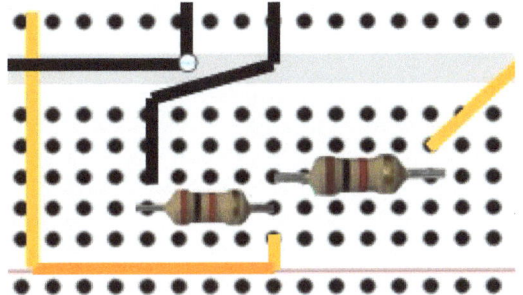

Using Logic Level Converters to Lower Voltage

A **"Logic Level Converter"** (Sometimes called a "Logic Level Shifter") is the way to go if you have very high speed (high baud rate) communications at 38,400 baud and above. A Logic Level Converter uses transistors to allow the higher (or lower) voltage to flow to the rest of the circuit.

A logic signal is simply a zero or a one, meaning it's either off or on. For this reason a transistor circuit can be used as a gate which controls the flow of a smaller amount of current. Normally a transistor uses a small current to control the base (the gate) that allows a larger current to flow between the collector (power source) and the emitter (going to ground), where the power source coming from the collector is higher than the power source controlling the gate. However in a Logic Level Converter (LLC) the power source is lower.

There are a couple ways that you can create a Logic Lever Converter:

1. MOSFET
2. Bi-directional Logic Level Shifter

Logic Level Conversion Using a MOSFET

FQP30N06L
N-Channel MOSFET
(www.fairchildsemi.com)

The FQP30N06L is a very popular MOSFTET. They are around 30 cents each on Ebay when bought in small quantities (5 or 10). They can handle up to 62 Volts and 30 Amps which can be managed by the 5 Volts that the Arduino digital output pins.

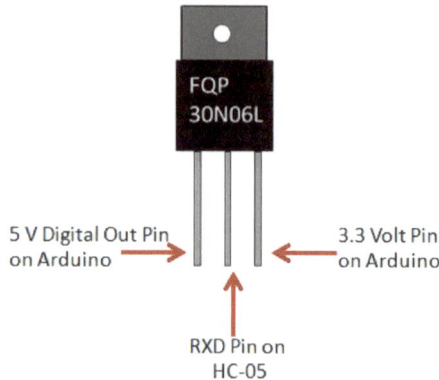

Logic Level Conversion Using a Bi-directional Logic Level Shifter

Bi-directional Logic Level Shifter
(www.Ebay.com)

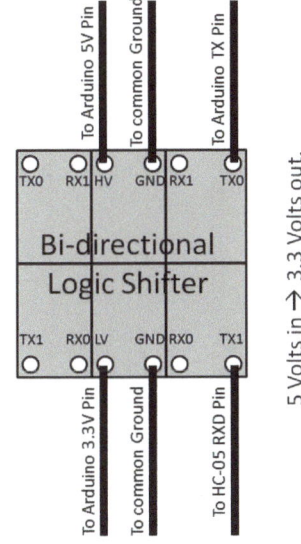

The **4-Channel Bi-directional Logic Level Shifter** is also called the "Quad Bi-directional Logic Level Converter" or just referred to as a "Quad LLC. It costs about $1 on Ebay without the pins attached and around $4-$5 fully assembled as you see in the picture above.

There are two sides, HV (High Voltage) and LV (Low Voltage). The labels are somewhat inconsistent, especially from the ones coming from China, but all function the same way.

The Low voltage side can handle 1.5 Volts – 7 Volts. The High voltage side can handle 1.5 Volts to 18 Volts. It is very simple to wire because it just goes in one side and out the other. No other components are needed. The representation shown here is only receiving one signal, the TX pin on the Arduino.

If you wanted to connect the RXD pin on the HC-05 to the RX1 pin on the LLC then the RX0 pin on the LLC to the Arduino RX pin, you could, but it is not necessary because the Arduino can accept 3.3 Volt signals without a problem.

I'm very happy with this type of Logic Level Converter. If you are converting logic signals to be used with **I2C devices**, then this would be the one to use because it is I2C friendly.

Other Multichannel Bi-Directional Logic Level Converters

8 Channel Bi-Directional
Logic Level Shifter
(www.Ebay.com)

There are other units out there. Some have eight channels such as the one shown to the left. It is a TXB0108 Eight Channel Bi-Directional Logic Level Converter/Shifter. made by Logic Gate Electronics. It costs around $5 or $2-$3 without the header pins attached. If you were building your own developer's breadboard, this would be a great one to include in your project.

Here is the pinout for the TXB0108:

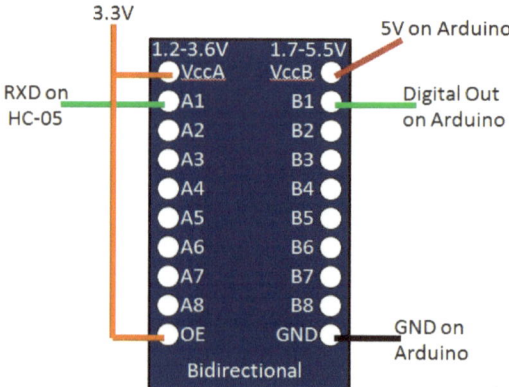

Since it is bi-directional, it can take a 5 Volt signal from the Arduino and convert it to a 3.3 volt signal for the HC-05 and take the 3.3 Volt signal and convert it up to a 5 Volts signal going the other way.

HC-05 AT Commands

AT	AT+DISC	AT+IPSCAN?	AT+PMSAD=	AT+SENM?
AT+ACDN?	AT+ENSNIFF=	AT+IPSCAN=	AT+POLAR?	AT+SENM=
AT+ADDR?	AT+EXSNIFF=	AT+LINK=	AT+POLAR=	AT+SNIFF?
AT+BIND?	AT+FSAD=	AT+MRAD?	AT+PSWD?	AT+SNIFF=
AT+BIND=	AT+IAC	AT+NAME?	AT+PSWD=	AT+UART?
AT+CLASS?	AT+IAC=	AT+NAME=	AT+RESET	AT+UART=
AT+CLASS=	AT+INQ	AT+ORGL	AT+RMAAD	AT+VERSION?
AT+CMODE?	AT+INQC	AT+PAIR=	AT+ROLE?	
AT+CMODE=	AT+INQM=	AT+PIO=	AT+ROLE=	

AT
Test command.
Typing AT by itself will make the module respond with OK of the UART serial connection is good.

AT+ADCN
Query the total number of authenticated devices from the authenticated device list.

AT+ADDR
Query the device's Bluetooth address.
It will return something like this:
+ADDR:14:3:5e875
When typing this address on another Bluetooth device in the role of master, you must convert the colons to commas, so it would be typed like this:
14,3,5e875

AT+BIND
Queries the address of the Bluetooth it is bound to. If there are none then it will return an address of 0:0:0.

AT+BIND=
Tells the module to bind to address of another Bluetooth module.
This command is only used if the AT+CMODE is set to 0 first.
For example:
AT+CMODE=0
AT+BIND=58AE,CD,9BD0E

AT+CLASS
Queries the class of the module.
The default is zero. This tells the Bluetooth module to broadcast to other modules what type of device it is. Generally this can be ignored and left at zero. However if you want to set it to a specific class, you can use the Bluetooth device class calculator found at:
http://www.ampedrftech.com/cod.htm

AT+CLASS=
Sets the Bluetooth class.
It tells other Bluetooth devices what type of device it is.
Zero is the default.
Example:
AT+CLASS=1

AT+CMODE
Check the connect mode.
The module will respond with the connection mode, but will drop the letter 'E' on the response. Just remember to include the 'E' when requesting or setting this value. For example, if the CMODE is set to zero and you request it, it will return the response:
+CMOD:0
Parameters:
0 – Connect only to bound devices
1 – Connect any slave device (default)
2 – Slave Loop

When mode 0 is used, the module will only allow another module with the specified address to link to it. All other requests from other Bluetooth devices will be ignored. If t his mode is used, then you have to use the AT+BIND command to set the specified address of the other module.

When mode 1 is used in Master mode, the module will connect to any slave device. However you must remember that only one other Bluetooth module can pair to your module at one time.

When mode 2 is used, it goes into a slave loop. This means that when a paired Bluetooth sends data, that data will be sent back to the sender. This would only be used if you are wanting to test the connection to see if any data is being lost or how fast the response would be.

AT+CMODE=

Sets the connection mode (0, 1 or 2).
Example:
AT+CMODE=0

AT+DISC

Disconnect.
Example:
AT+DISC
The SPP must be initialized first using the AT+INIT command before you can use the AT+DISC command.

AT+ENSNIFF=

Enter energy saving mode when connected with another device.
Example:
AT+ENSNIFF=14,3,53875

AT+EXSNIFF=

Exits energy saving mode.

AT+FSAD=

Search authenticated device
Example:
AT+FSAD=14,3,53875

AT+IAC

Check the General Inquirer Access Code.
This will return the default value like so:
+IAC:9e8b33

AT+IAC=

Sets the General Inquirer Access Code.
There are only two values that can be used here:
AT+IAC= 9e8b33 (the default)
or
AT+IAC=9e8b00

AT+INQ

Searches for another Bluetooth device.
It will return with a list of the devices.
It has three parameters:
Parameter #1 = Address
Parameter #2 = Class
Parameter #3 = Signal Strength
The device must be set as a master with the AT+ROLE=1 command.
The AT+INIT command must be set before the AT+INQ command
is used.

Example:
AT+ROLE=1
AT+INIT
AT+INQ

If there is another Bluetooth device in range it will return with
something like:
+INQ:14:3:5e875,1F1F,FFBA

The address of the device would be 14:3,5e875 but when connecting to
it you would replace the colons with commas like so:

AT+LINK=14,3,5e875

AT+INQC
Cancel the inquiring of other Bluetooth devices.

AT+INQM
Check the inquiry access patterns.
Organizes the list of nearby devices found during an inquiry mode.
There are three parameters:
Parameter #1 = Inquiry mode (0 – standard, 1 – by signal strength)
Parameter #2 = Max number of Bluetooth devices to respond to
Parameter #3 = Timeout (1-48)
Example return:
+INQM:1,1,48

AT+INQM=
Sets the inquiry access patterns.
For example, set standard mode, stop at 9devices and search for 5 seconds:
AT+INQM=0,9,5

AT+IPSCAN
Checks the scan parameter
There are four parameters:
Parameter #1 = Query time interval
Parameter #2 = Query duration
Parameter #3 = Paging interval
Parmeter #4 = Call duration
The default is 1024,512,1024,512

AT+IPSCAN=
Sets the scan parameters.
Example:
AT+IPSCAN=1000,500,1200,250

AT+LINK=

Connect to another Bluetooth device.
This is the command you must use to make the connection to another Bluetooth device. The other device is automatically paired once its address set with the AT+BIND= command.
Example:
AT+LINK=14,3,5e875

AT+MRAD

Displays the most recently used authenticated Bluetooth device.
An example output would be:
+MRAD:2016:6:61551

AT+NAME

Gets the name of the device.
Example:
+NAME:MYROBOT
The default name is H-C-2010-06-0, depending up on the version.

AT+NAME=

Sets the name of the device.
Example:
AT+NAME=MYROBOT

AT+ORGL

Restores the Module to its factory preset default settings:
CLASS: 0
NAME: H-C-2010-06-01
IAC: 9e8b33
ROLE: 0 (Slave mode)
PSWD: 1234
INQM: 1,9,48
UART: 38400,0,0 (38,400 baud, 1 stop bit, no parity)
POLAR: 1,1

AT+PAIR=

Add another Bluetooth Device to the list of devices that can be paired with.

It has two parameters:
Parameter #1 = Address of the remote device
Parameter #2 = Timeout in seconds
Example:
AT+PAIR=14,3,5e875,10 //responds with ok if successful within 10 seconds

AT+BIND=14,3,5e875 //remember this paired device
AT+CMODE=1   //set so module can only connect with devices in the pair list (optional)

AT+LINK=14,3,5e875 //Links to the address of the device.

Now these two devices will automatically be paired when they are turned on.
After LINK command is typed, the module should switch into communications mode. The light should stop blinking at two second intervals.

AT+PIO=
Set the output as High or Low for a port.
This has two parameters:
Parameter #1 = Port #
Parameter #2 = Port Status
Example:
AT+PIO=10,1

AT+PMSAD=
Delete an authenticated device in the pair list.
Example:
AT+PMSAD=14,3,5e875

AT+POLAR
Gets the module's drive indicator and connection status LED.
Output example:
+POLAR:1,1

AT+POLAR=

Sets the module's drive indicator and connection status LED.
There are two parameters.
Parameter #1 is for the PI08. 0 tells it to output low level and turn on the LED. 1 tells it to output high level and ton on the LED.
Parameter #2 is for the PI09. 0 tells it to output low level and indicate a successful connection. 1 tells it to output a high level and indicate a successful connection.
The default is 1,1

AT+PSWD

Displays the password.
Example:
PSWD:1234

AT+PSWD=

Sets the password.
Example:
AT+PSWD=5678

AT+RESET

Returns the module to communications mode 38400 baud.
Resets the module and exits AT mode.
This command is normally given after changing the role of the module.
(See AT+ROLE=).

AT+RMAAD

Deletes all authenticated Bluetooth devices.

AT+ROLE

Get the role of the module.
0=Slave

1=Master
2=Slave Loop

AT+ROLE=

Sets the role of the module.
For example, to set it to master mode:
AT+ROLE=1
The values can be 0, 1 or 2.

AT+SENM

Check the security mode.
There are two parameters (#1 Security mode and #2 HCI Enc mode)
The default is 0,0
The values for parameter #1 may be 0,1,2,3 or 4.
The values for parameter #2 may be 0,1 or 2.

AT+SENM=

Set the security mode.
Example:
AT+SENM=1,1

AT+SNIFF

Get the energy saving parameter.
This has to do with power consumption during sleep time for the module.
It has four parameters:
Parameter #1 = Maximum time
Parameter #2 = Minimum time
Parameter #3 = Test time
Parameter #4 = Limited time
The default is 0,0,0,0

AT+SNIFF=

Sets the energy savings parameters:
Example:
AT+SNIMM=5, 1, 3, 0

This isn't a feature that is normally used. It doesn't affect the transmission of data.

AT+UART

Inquire the serial parameters.
There are three parameters.
Parameter #1 = Baud Rate (4800, 9600, 19200, 38400, 57600, 115200, 234000,460800, 921600, 1382400)
Parameter #2 = Stop Bit (0=1 bit, 1= 2 bits)
Parameter #3 = Parity Bit (0=none, 1=Odd, 2=Even)
The default is:
38400, 0, 0 (38,400 Baud, 1 stop bit, No parity)

AT+UART=

Sets the serial parameters
Example:
AT+UART=38400,0,0
The default is 38400,0,0

AT+VERSION

Gets the version number.
Example:
+VERSION:2.0-20100601

Error Codes

0	Command Error/Invalid Command
1	Results in default value
2	PSKEY write error
3	Device name is too long (>32 characters)
4	No device name specified (0 length)
5	Bluetooth address NAP is too long
6	Bluetooth address UAP is too long
7	Bluetooth address LAP is too long
8	PIO map not specified (0 length)
9	Invalid PIO port Number entered
A	Device Class not specified (0 length)
B	Device Class too long
C	Inquire Access Code not Specified (0 length)
D	Inquire Access Code too long
E	Invalid Iquire Access Code entered
F	Pairing Password not specified (0 length)
10	Pairing Password too long (> 16 characters)
11	Invalid Role entered
12	Invalid Baud Rate entered
13	Invalid Stop Bit entered
14	Invalid Parity Bit entered
15	No device in the Pairing List
16	SPP not initialized
17	SPP already initialized
18	Invalid Inquiry Mode
19	Inquiry Timeout occured
1A	Invalid/zero lenght address entered
1B	Invalid Security Mode entered
1C	Invalid Encryption Mode entered

Standard Resistor Values

Carbon Film, ¼ Watt, 5% Tolerance

1	20	300	3.9K	51K	680K
1.5	24	330	4.3K	56K	750K
1.8	27	360	4.7K	62K	820K
2	30	390	5.1K	68K	910K
2.2	33	430	5.6K	75K	1M
2.4	36	470	6.2K	82K	1.1M
2.7	39	510	6.8K	91K	1.2M
3	43	560	7.5K	100K	1.3M
3.3	47	620	8.2K	110K	1.5M
3.6	51	680	9.1K	120K	1.6M
3.9	56	750	10K	130K	1.8M
4.3	62	820	11K	150K	2M
4.7	68	910	12K	160K	2.2M
5.1	75	1K	13K	180K	2.4M
5.6	82	1.1K	15K	200K	2.7M
6.2	91	1.2K	16K	220K	3M
6.8	100	1.3K	18K	240K	3.3M
7.5	110	1.5K	20K	270K	4.7M
8.2	120	1.6K	22K	300K	5.1M
9.1	130	1.8K	24K	330K	5.6M
10	150	2K	27K	360K	6.2M
11	160	2.2K	30K	390K	6.8M
12	180	2.4K	33K	430K	7.5M
13	200	2.7K	36K	470K	8.2M
15	220	3.0K	39K	510K	9.1M
16	240	3.3K	43K	560K	10M
18	270	3.6K	47K	620K